Logic Puzzles For Kids Ages 6-8

This book belongs to:

_____

Thank you for choosing our logic puzzle book. It's great that you like logic puzzles as much as we do! These activities offer hours of fun and are a great way to enhance your brain functions and critical thinking skills.

There are all kinds of logic puzzles in this book. We've organized it from easy to hard difficulty levels!

Once you complete the book, there will be a nice gift waiting for you as your reward.

Have fun and enjoy!

Level 1

Let's start! In this level, you will go through easy puzzles.

Ready? Let's go!

Match the eggs with the right basket size!

3

How many of each shape are there?

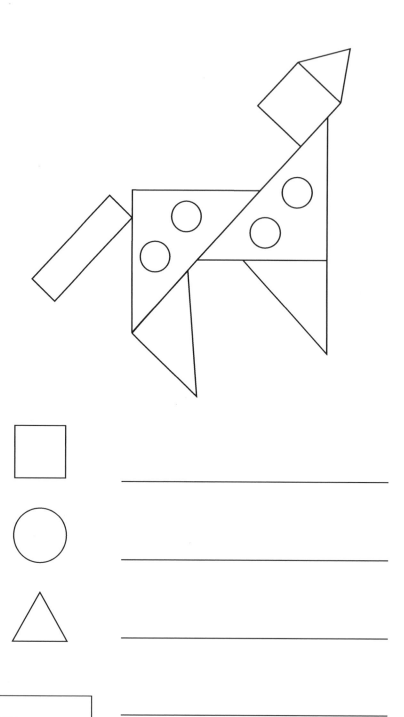

⬜ _____

⭕ _____

🔺 _____

▭ _____

How many fishes are in the tank?

write the answer in the box

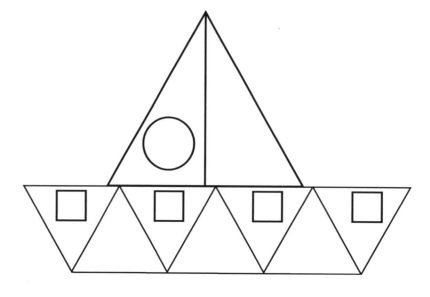

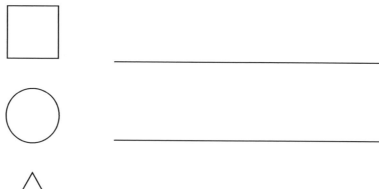

Count the coins!

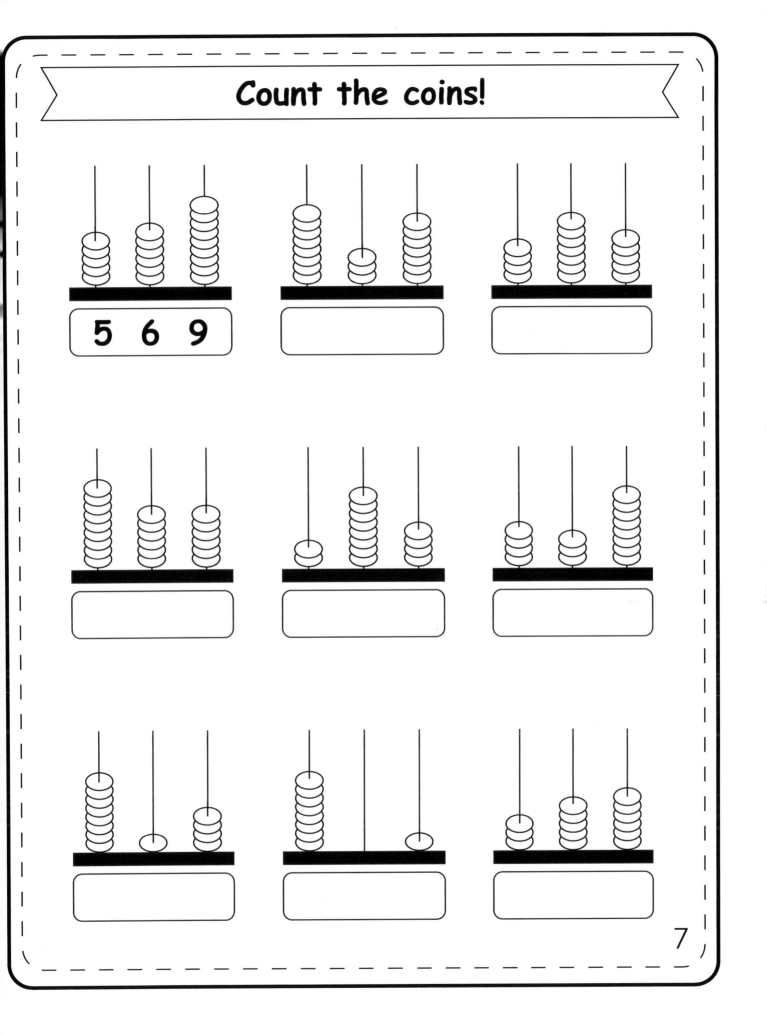

8

Pick the correct path to catch the thief!

Write your answer

How many of each shape are there?

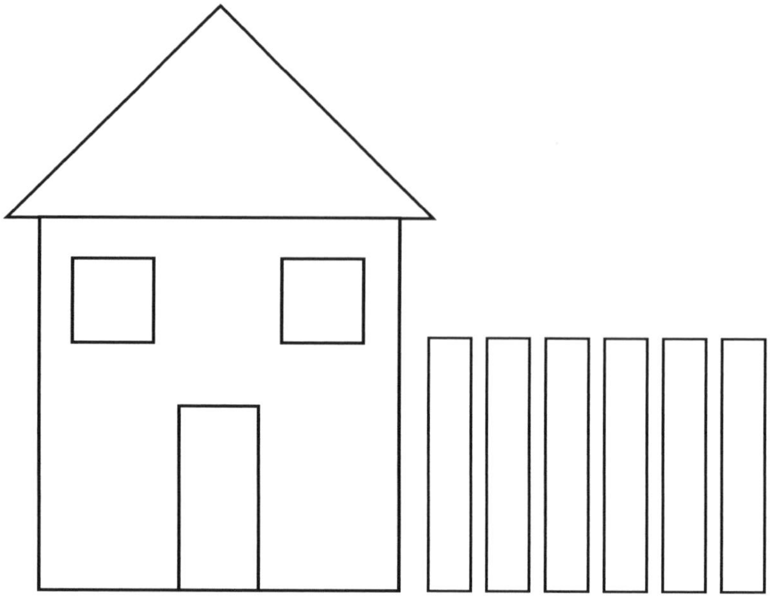

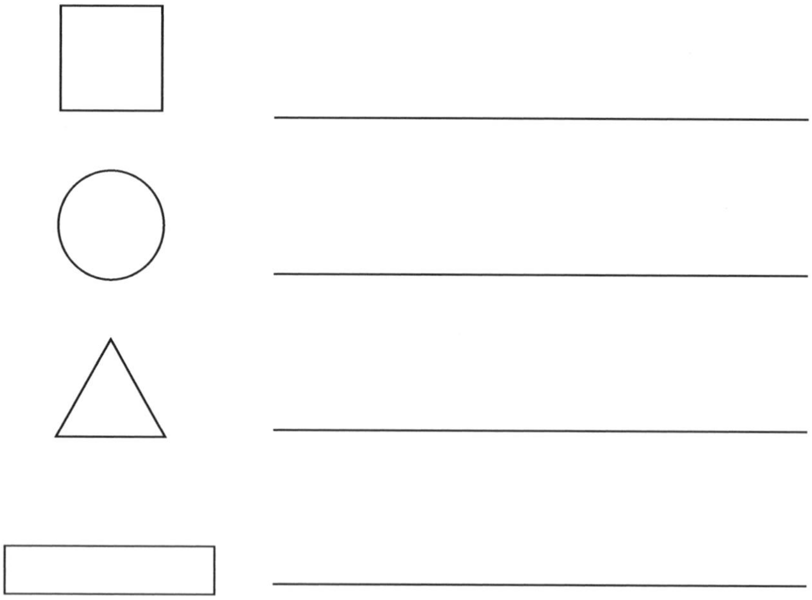

Trick Question: How old is everyone?

- Kelly is 3 years older than Peter.

- Jack is the same age as Ann.

- Max is 2 years older than Kelly.

- Peter is 9 years old.

- Ann is 3 years younger than Max and 2 years older than Defny.

Peter

Kelly

Max

Ann

Jack

Defny

Name that animal!

H S F
I

☐ ☐ ☐ ☐

R
T
I B
B A

☐ ☐ ☐ ☐ ☐ ☐

E
F R I E F
G I A

☐ ☐ ☐ ☐ ☐ ☐ ☐

N A I S
A L

☐ ☐ ☐ ☐ ☐

What day was it yesterday? And what day is it tomorrow?

YESTERDAY	TODAY	TOMORROW
	Wednesday	
	Thursday	
	Monday	
	Saturday	
	Tuesday	
	Friday	
	Sunday	

Wednesday

Monday   Sunday

Tuesday   Thursday

Saturday   Friday

13

Find & write the correct answer in the box

$$\text{hat} + \text{hat} = 20$$

$$\text{violin} + \text{watch} = 80$$

$$\text{watch} + \text{hat} = 40$$

$$\text{hat} + \text{violin} + \text{watch} = \boxed{}$$

?

14

Match with the shadow

Circle the shape that cannot be made with the lines on the left. The first is completed.

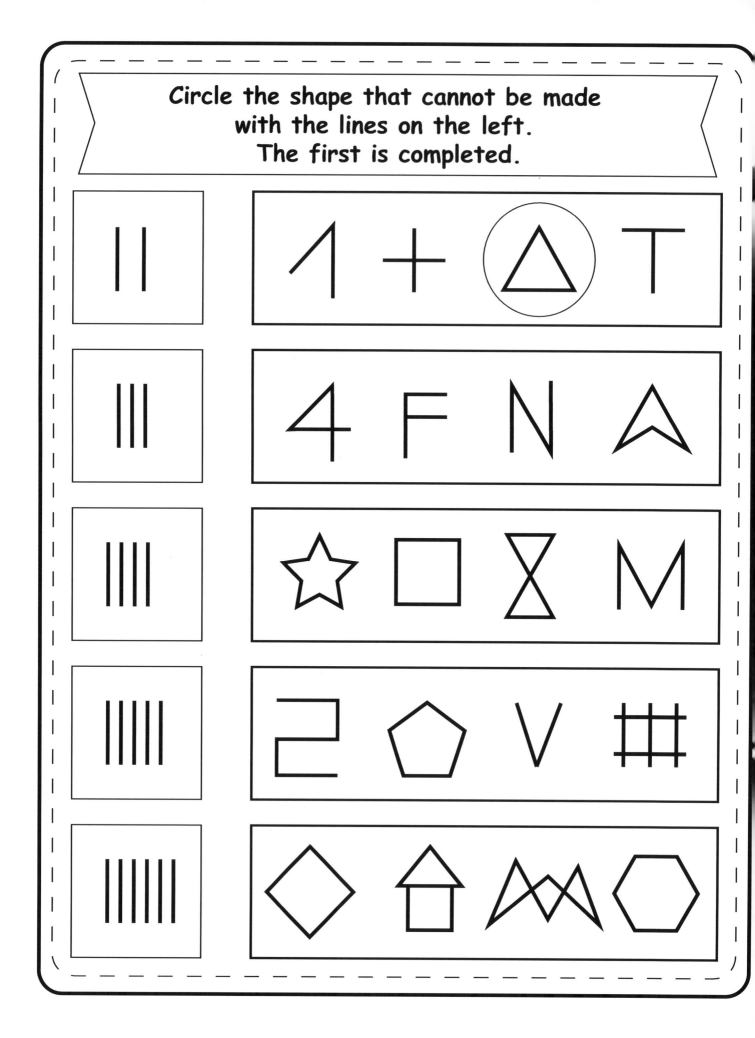

Level 2

In this level, you will go through medium difficulty puzzles.

Ready? Let's go!

Can you figure out where each shape is located on the grid?

	A	B	C	D	E	F	G
7		★					
6						●	
5	▮						
4		➤					⬟
3				■			
2		▲					
1							

★	B,7
■	
▲	
▮	
➤	
●	

Draw each shape in its correct position on the grid!

★	B,4
■	A,1
▲	G,3
▮	F,7
➤	C,6
●	D,2

	A	B	C	D	E	F	G
7							
6							
5							
4							
3							
2							
1							

Multiple Choice:
Choose the right answer!

a.

b.

c.

d.

🫖 + 🫖 = **26**

11 + 🍾 = ☕

🍾 + 🫖 = **17**

☕ - 🫖 = **3**

☕ = **?**

Write your answer

Count the coins and color them in!

6 2 8

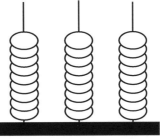

7 3 2

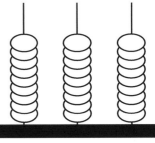

8 0 4

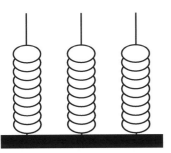

3 5 7

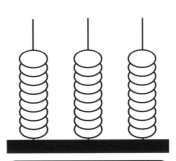

4 2 0

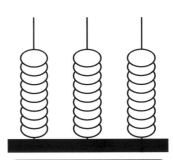

2 6 3

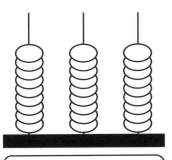

5 8 8

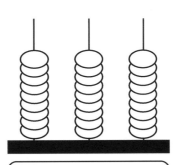

1 9 2

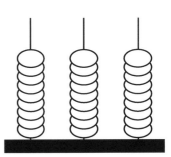

9 7 4

Multiple Choice: Choose the right directions!

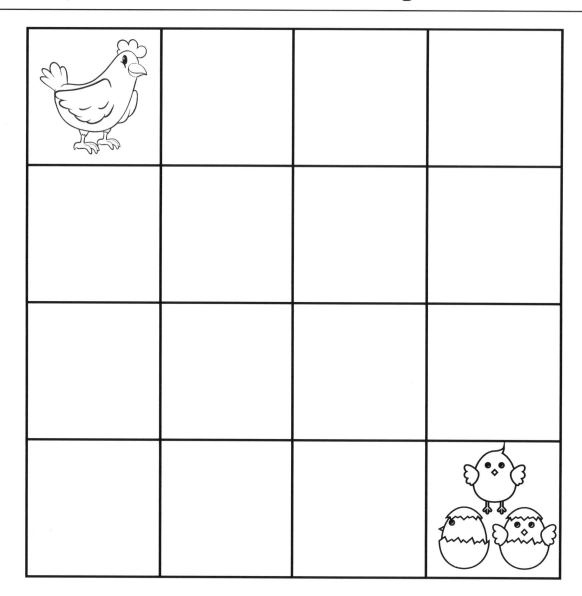

Select & underline the correct way

01) → → ↓ ↓ →

02) ↓ ↓ └→ → →

03) → ↓ ↓ ↓ ↓

Match the shadow

Count them up!

How many of each fruit are there?

HELP THE ASTRONAUT

GET BACK TO HIS SHIP →

Count them up!

Spelling Test: Find out what uncommon letter each word has!

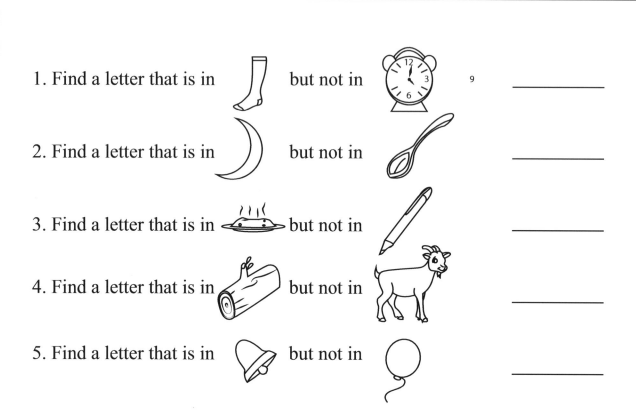

1. Find a letter that is in [sock] but not in [clock] 9 _____

2. Find a letter that is in [moon] but not in [spoon] _____

3. Find a letter that is in [pie] but not in [pen] _____

4. Find a letter that is in [log] but not in [goat] _____

5. Find a letter that is in [bell] but not in [balloon] _____

Write your letters again here to spell a happy word

_____ _____ _____ _____ _____

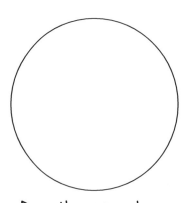

Draw the answer here

Count the school supplies!

How many?

Match the shadow

Tricky Math: Can you figure out this math problem?

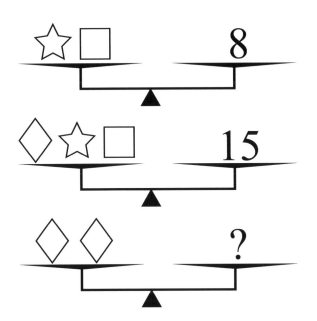

☆ □   8

◇ ☆ □   15

◇ ◇   ?

☆ □ = _____

◇ = _____

? = _____

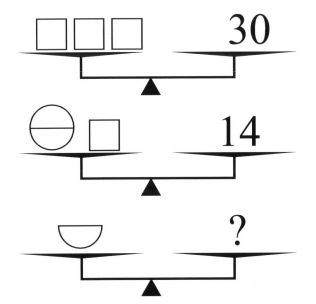

□ □ □   30

⊖ □   14

⌓   ?

□ = _____

⊖ = _____

? = _____

31

Match with the missing half

Find & write the correct answer in the box

🐕 + 🚢 = **7**

🚀 + 🐕 + 🚢 = **17**

🚢 − 🚀 = **3**

🚀 × 🚢 × 🐕 = **?**

Write your answer

Level 3

Wow you've made it far!

In this level, you will go through more difficult puzzles.

Ready? Let's go!

Connect all the dots with 4 straight lines without retracing them

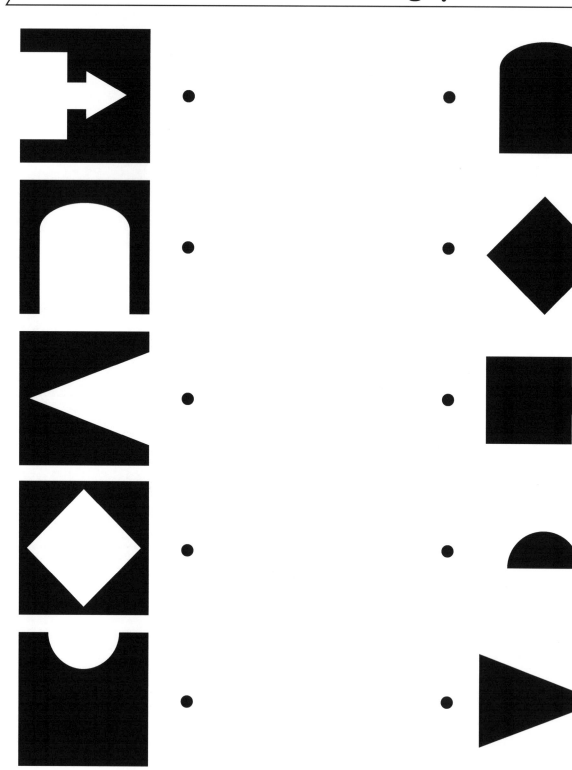

Fill in the boxes with the correct numbers

$$4 - 2 = \boxed{}$$

$$+$$

$$=$$

$$\boxed{} - 7 = 2$$

$$\boxed{} - 1 =$$

$$+ 8 = \boxed{}$$

$$+$$

$$-$$

$$+ 6 = \boxed{}$$

$$6$$

$$=$$

$$-$$

$$=$$

$$\boxed{} - 5 = \boxed{}$$

$$- \boxed{} = 5$$

$$-$$

$$+$$

$$=$$

$$1$$

$$3 - \boxed{} =$$

$$=$$

$$=$$

$$+$$

$$8$$

$$9 - \boxed{} = \boxed{}$$

$$5$$

$$-$$

$$+$$

$$=$$

$$5$$

$$=$$

$$3 + 3 = \boxed{}$$

$$\boxed{} - 4 = \boxed{}$$

Find the words given below

C	A	N	M
A	M	D	S
T	A	N	A
B	P	A	T

CAT   PAT

BAD   TAN

MAP   SAT

MAN   NAP

MAD   MAT

PAD   TAD

M	M	P	P
A	A	M	A
N	T	A	D
M	A	D	N

C	P	A	G
A	R	A	N
B	B	A	A
T	A	P	B

BAG   RAN

CAB   CRAB

NAB   TAP

In what parking spot number is the car parked?

16   06   68   88   98

Count the letters!

p	q	q	b	d	p	q	b
d	q	p	b	d	p	q	d
d	b	p	q	b	p	d	d
b	p	q	q	p	d	b	b
d	p	b	q	q	p	b	d
p	b	q	q	p	p	d	b
d	p	b	d	q	p	d	b

b _____   d _____

p _____   q _____

Fill in the boxes with the correct numbers

$3 + 2 = \square$ $\square + 4 = 8$

$+$ $+$

1 5 6

$=$ $=$ $+$

3 $\square + \square = \square$ 2

$+$ $+$ $=$

2 $\square + \square = \square$

$=$ $=$ $+$

$\square + \square = 7$ $\square + \square = 7$

$+$ $=$ $+$

$6 + \square = 9$ 4

$=$ $=$

$6 + \square = 8$ $\square + 3 = \square$

$+$ $+$

2 5

$=$ $=$

$\square + \square = \square$

41

How many of each shape are there?

Match with the shadow

43

Match with the missing half

Count Them Up: Can you find all the drawings on this picture?

Match with the shadow

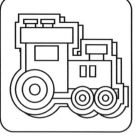 

Count Them Up!

[leaf] = [ ]

[chicken] = [ ]

[hen] = [ ]

[chick in nest] = [ ]

[chick] = [ ]

Find & write the answer

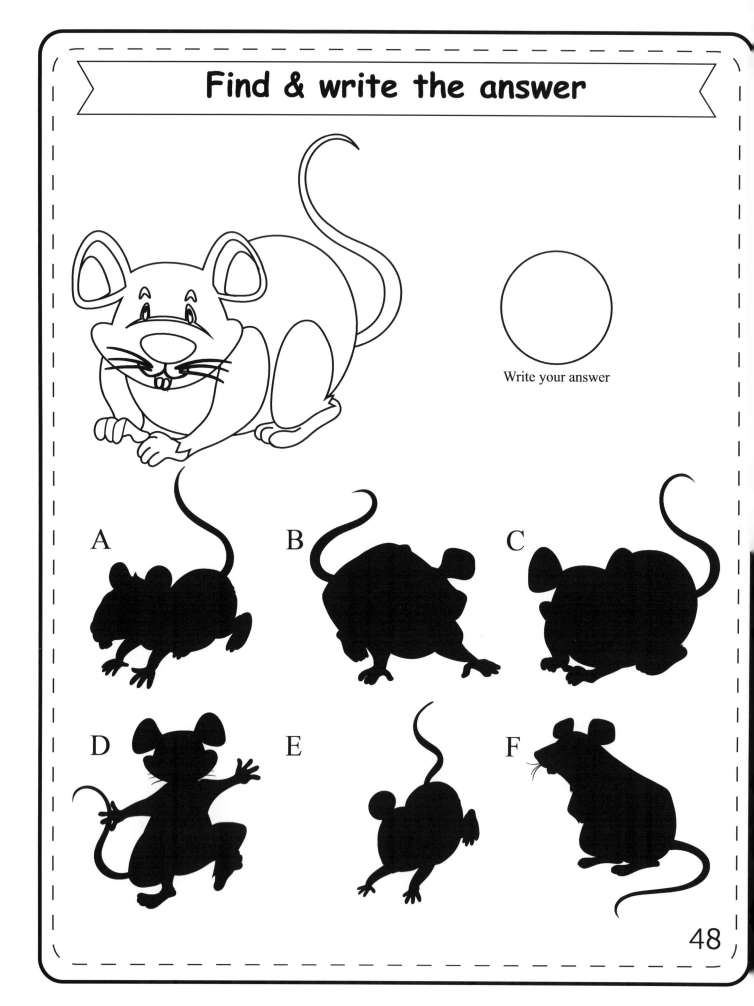

Write your answer

A   B   C

D   E   F

48

Match with the missing half

49

Copy the shaded circles and count how many circles you shaded in!

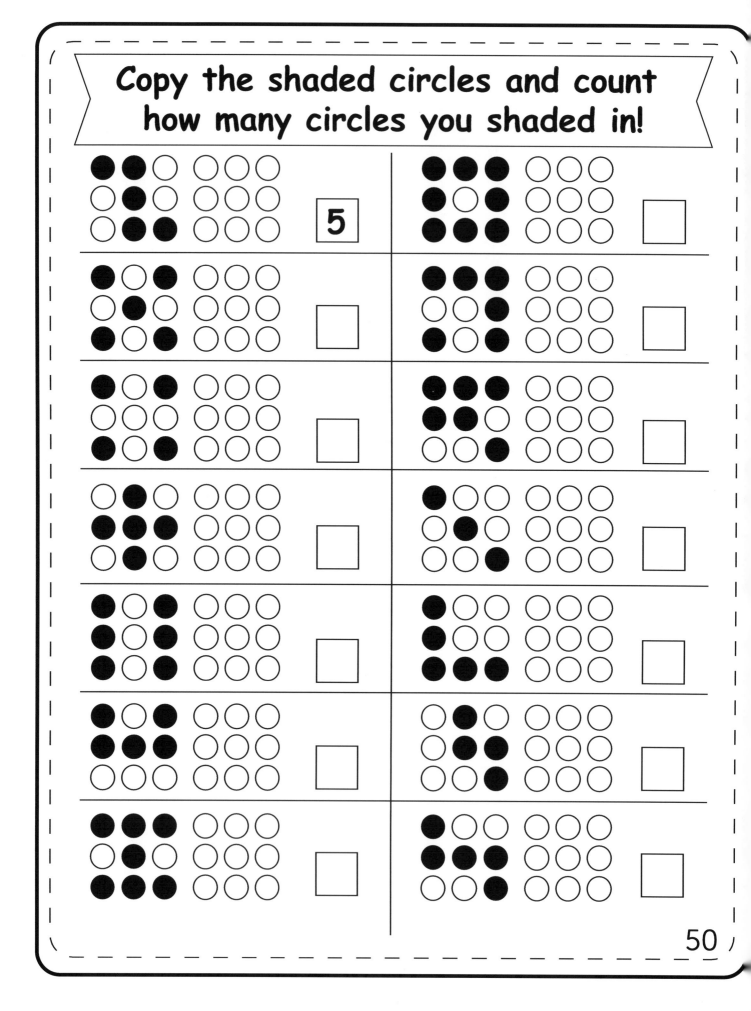

$7 - 3 = \boxed{}$

$\boxed{} - 2 = \boxed{}$

$+\ 5\ =$ (down from top-left result)

$-\ 3\ =$ (down from top-right)

$\boxed{} + 1 = \boxed{}$

$3 + \boxed{} = \boxed{}$

$-\ 2$ (down)

$-$ (down)

$\boxed{} + 4 = \boxed{}$

$=$ ... 5

$-\ 3\ =$

$\boxed{} - 2 = \boxed{}$

$+\ 3\ =$ (down)

$\boxed{} + 4 = \boxed{}$

$+$ (down)

$2 + \boxed{} = \boxed{}$

$\boxed{} - 5 = \boxed{}$

$-\ 3\ =$ (down)

$-\ 3\ =$ (down)

$=$

$3 + \boxed{} = 5$

$\boxed{} + 4 = \boxed{}$

How many of each shape are there?

○ __ □ __ △ __

Solve this!

+ + = **60**

+ + = **30**

− = **3**

+ + = ☐ **?**

3

Match the Eggs With the Right Basket Size!

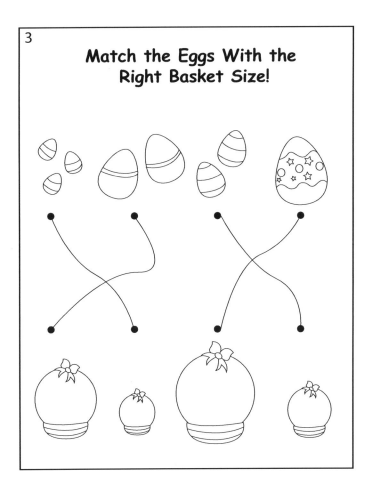

4

How many of each shape are there?

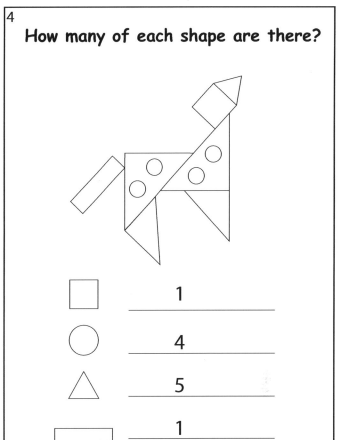

⬜	1
◯	4
△	5
▭	1

5

How many fishes are in the tank?

31

write answer in the box

6

How many of each shape are there?

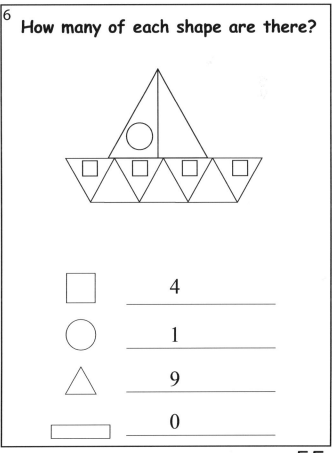

⬜	4
◯	1
△	9
▭	0

55

7

Count the coins!

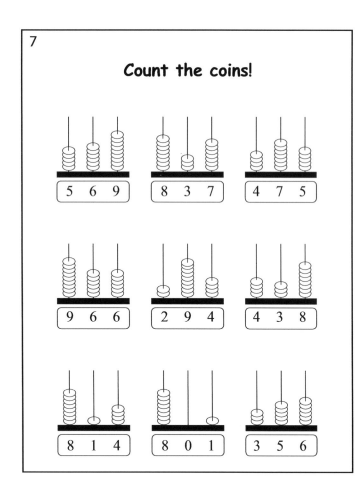

5	6	9
8	3	7
4	7	5

9	6	6
2	9	4
4	3	8

8	1	4
8	0	1
3	5	6

8

Match the tower to the right blocks!

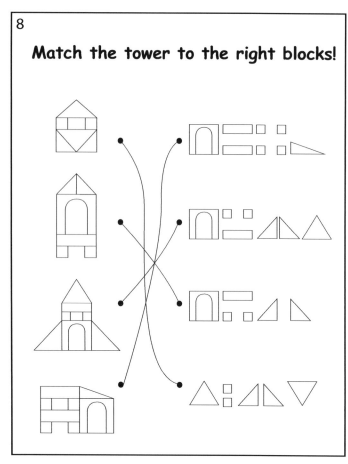

9

Pick the correct path to catch the thief!

D

Write your answer

10

How many of each shape are there?

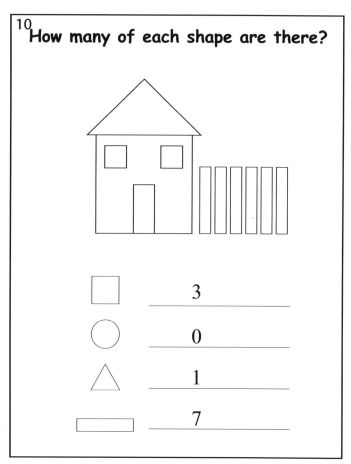

☐	3
◯	0
△	1
▭	7

56

11

Trick Question: How old is everyone?

- Kelly is 3 years older than Peter.
- Jack is the same age as Ann.
- Max is 2 years older than Kelly.
- Peter is 9 years old.
- Ann is 3 years younger than Max and 2 years older than Defny.

12

Name that animal!

F I S H

R A B B I T

G I R A F F E

S N A I L

13

What day was it yesterday? And what day is it tomorrow?

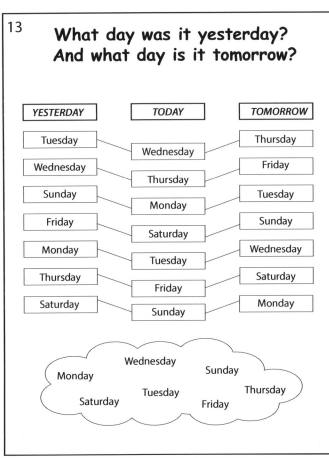

YESTERDAY	TODAY	TOMORROW
Tuesday	Wednesday	Thursday
Wednesday	Thursday	Friday
Sunday	Monday	Tuesday
Friday	Saturday	Sunday
Monday	Tuesday	Wednesday
Thursday	Friday	Saturday
Saturday	Sunday	Monday

Wednesday
Monday   Sunday
Tuesday   Thursday
Saturday   Friday

14

Find & write the correct answer in the box

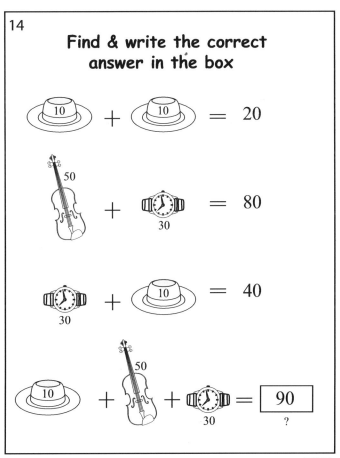

90

15
Match with the shadow

16
Circle the shape that cannot be made with the lines on the left. The first is completed

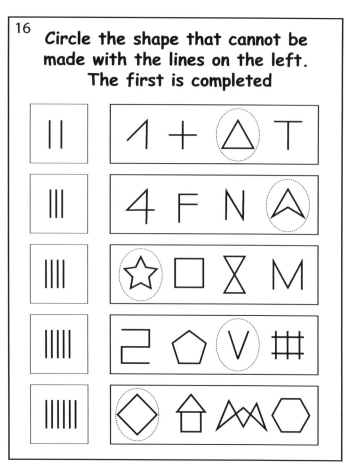

17
Can you figure out where each shape is located on the grid?

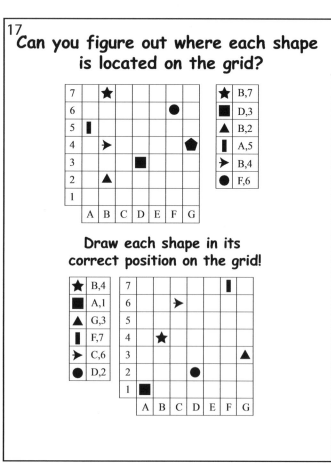

Draw each shape in its correct position on the grid!

19
Multiple Choice: Choose the right answer!

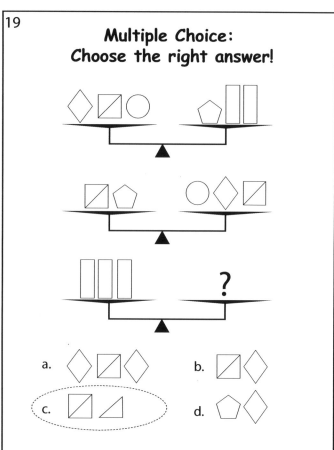

20

Find & write correct answer in the box

$\text{kettle}_{13} + \text{kettle}_{13} = \mathbf{26}$

$\mathbf{11} + \text{bottle}_{05} = \text{cup}_{16}$

$\text{bottle}_{05} + \text{kettle}_{13} = \mathbf{17}$

$\text{cup}_{16} - \text{kettle}_{13} = \mathbf{3}$

$\text{cup}_{16} = ?$

16

Write your answer

21

Count the coins and color them in!

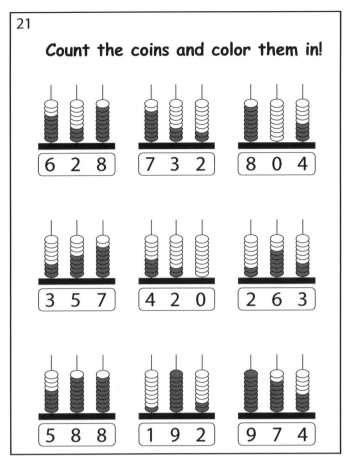

6	2	8

7	3	2

8	0	4

3	5	7

4	2	0

2	6	3

5	8	8

1	9	2

9	7	4

22

Multiple Choice: Choose the right directions!

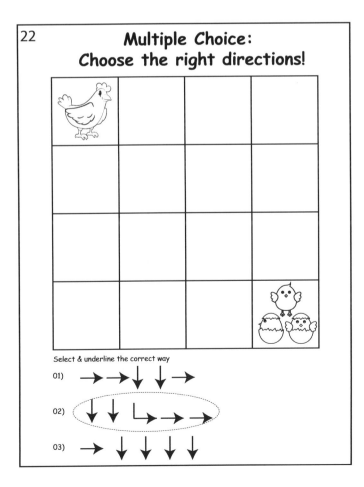

Select & underline the correct way

01) → → ↓ ↓ →

02) ↓ ↓ ↳ → →

03) → ↓ ↓ ↓ ↓

23

Match the shadow

24

Count them up!

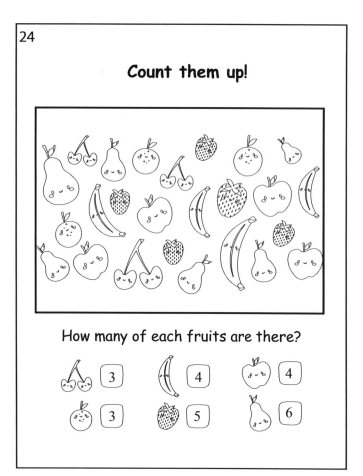

How many of each fruits are there?

🍒 3   🍌 4   🍎 4

🍊 3   🍓 5   🍐 6

25

← HELP THE ASTRONAUT

GET BACK TO HIS SHIP →

26

Guess the Crossword Puzzle!

27

Count them up!

⚀ 7   ⚅ 6

⚂ 10   ⚄ 5

28

Spelling Test: Find out what uncommon letter each word has!

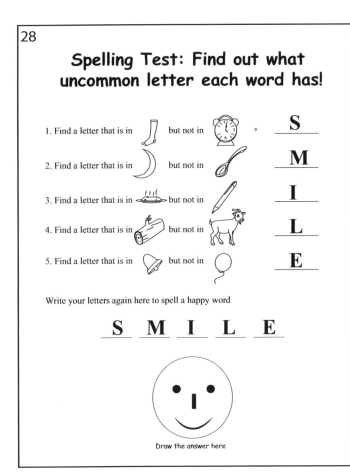

1. Find a letter that is in [sock] but not in [clock] 9   S

2. Find a letter that is in [moon] but not in [spoon]   M

3. Find a letter that is in [pie] but not in [pen]   I

4. Find a letter that is in [log] but not in [goat]   L

5. Find a letter that is in [bell] but not in [balloon]   E

Write your letters again here to spell a happy word

S M I L E

Draw the answer here

29

Count the school supplies!

How many?

book	3	brush	3	apple	4
eraser	4	bulb	2	crayon	3

30

Match the shadow

31

Tricky Math: Can you figure out this math problem?

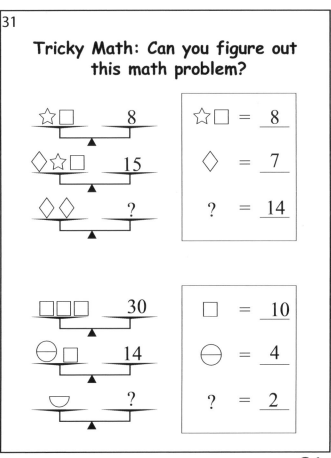

☆□ = 8
◇ = 7
? = 14

□ = 10
⊖ = 4
? = 2

32

Match with the missing half

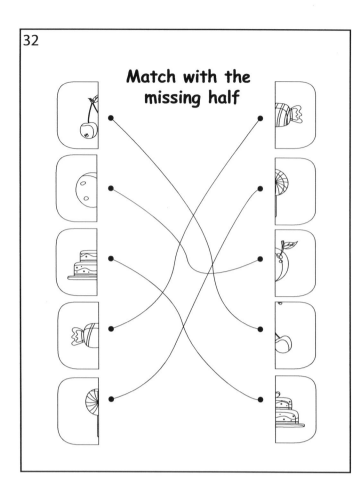

33

Find & write the correct answer in the box

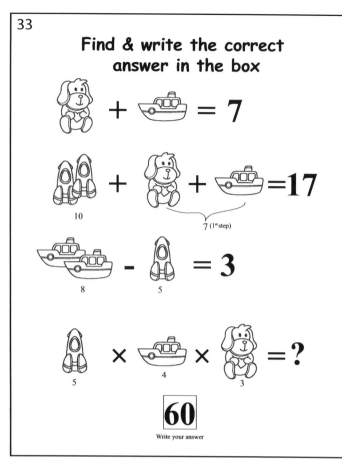

60
Write your answer

35

Connect all the dots with 4 straight lines without retracing them

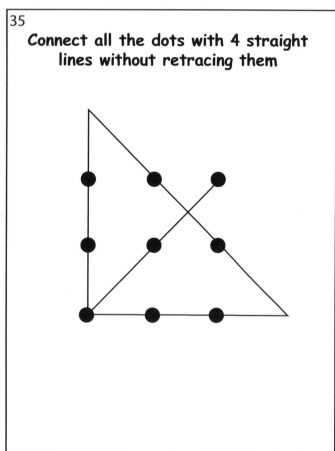

36

Match the missing part

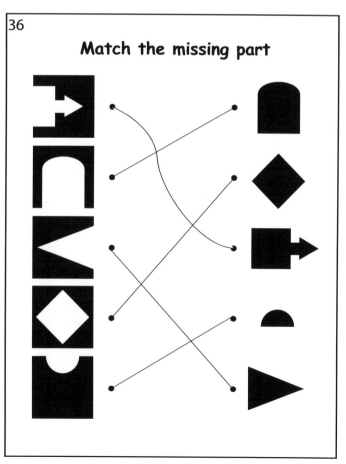

37 Fill in the boxes with the correct numbers

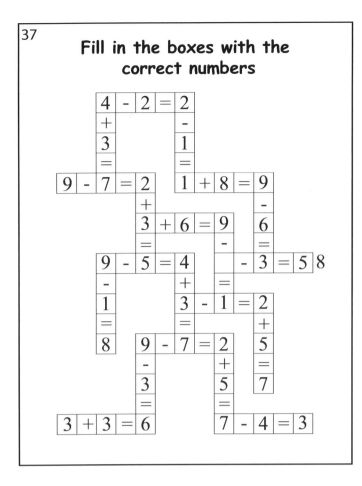

				4	-	2	=	2			
				+				-			
				3				1			
				=				=			
9	-	7	=	2		1	+	8	=	9
				+				-	
				3	+	6	=	9		6
				=				-		=
9	-	5	=	4			-	3	=	5	8
-				+		=	
1				3	-	1	=	2
=				=				+
8		9	-	7	=	2		5
		-				+		=
		3				5		7
		=				=	
3	+	3	=	6		7	-	4	=	3

38 Find the words given below

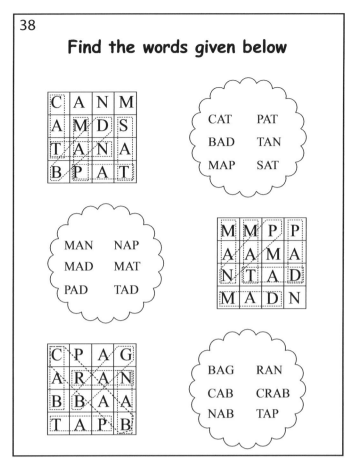

CAT   PAT
BAD   TAN
MAP   SAT

MAN   NAP
MAD   MAT
PAD   TAD

BAG   RAN
CAB   CRAB
NAB   TAP

39 In what parking spot number is the car parked?

The answer is 87, look at the image upside down:

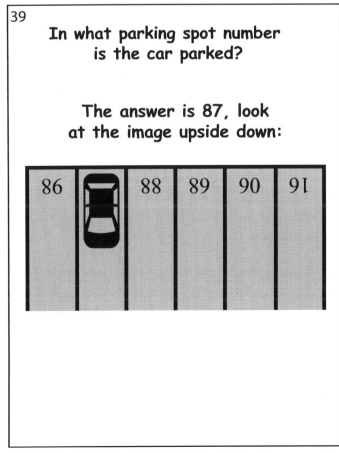

40 Count the letters!

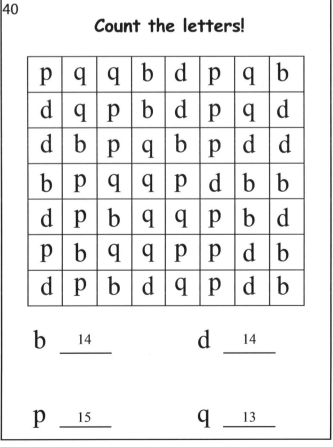

p	q	q	b	d	p	q	b
d	q	p	b	d	p	q	d
d	b	p	q	b	p	d	d
b	p	q	q	p	d	b	b
d	p	b	q	q	p	b	d
p	b	q	q	p	p	d	b
d	p	b	d	q	p	d	b

b ___14___   d ___14___

p ___15___   q ___13___

63

41 Fill in the boxes with the correct numbers

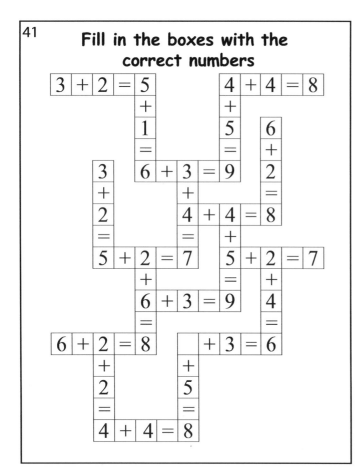

$3 + 2 = 5 \qquad 4 + 4 = 8$

$+ \qquad +$

$1 \qquad 5 \qquad 6$

$= \qquad = \qquad +$

$3 \qquad 6 + 3 = 9 \qquad 2$

$+ \qquad + \qquad =$

$2 \qquad 4 + 4 = 8$

$= \qquad = \qquad +$

$5 + 2 = 7 \qquad 5 + 2 = 7$

$+ \qquad = \qquad +$

$6 + 3 = 9 \qquad 4$

$= \qquad =$

$6 + 2 = 8 \qquad \_ + 3 = 6$

$+ \qquad +$

$2 \qquad 5$

$= \qquad =$

$4 + 4 = 8$

42 How many of each shape are there?

▢	5
◯	7
△	7
▭	4

▢	2
◯	17
△	7
▭	5

B

▢	2
◯	12
△	3
▭	2

43 Match with the shadow

44 Match with the missing half

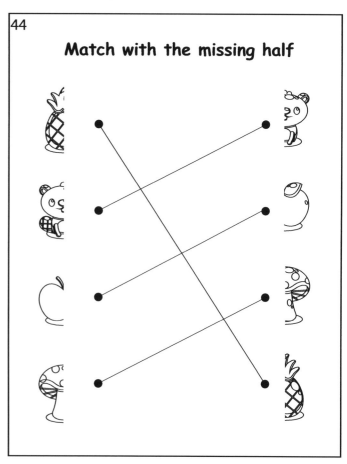

45

Count Them Up: Can you find all the drawings on this picture?

| 5 | 1 | 3 | 3 | 9 | 2 | 6 |

46

Match with the shadow

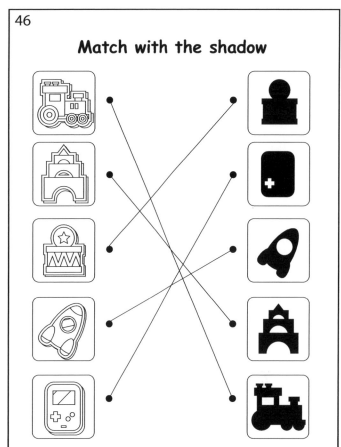

47

Count Them Up!

48

Find & write the answer

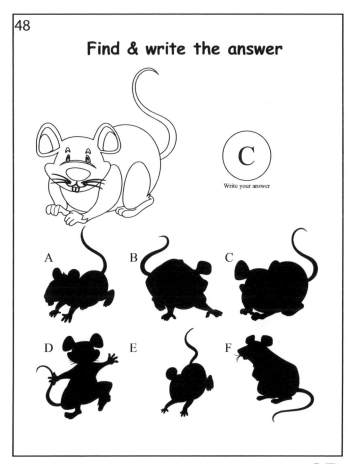

C

Write your answer

Match with the missing half

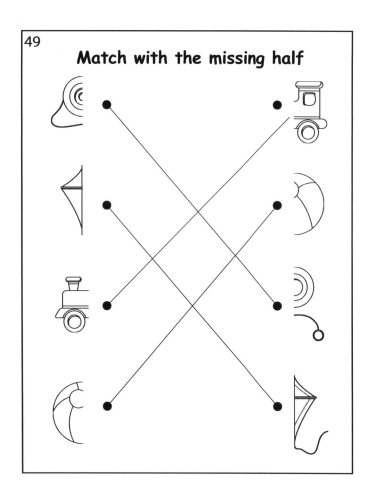

Copy the shaded circles and count how many circles you shaded in!

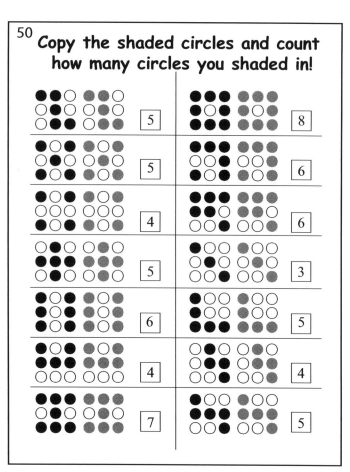

Fill in the boxes with the correct numbers

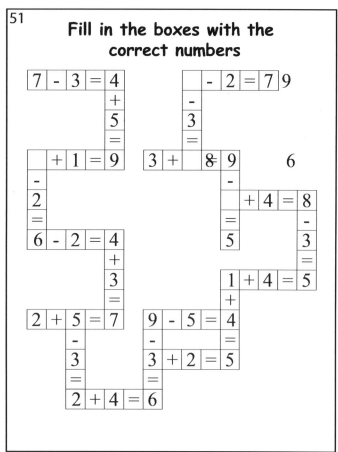

7 - 3 = 4

- 2 = 7 9

+ 5 =

- 3 =

+ 1 = 9   3 + 8 9   6

-2

=   + 4 = 8

6 - 2 = 4   5   -

+3   3

=   1 + 4 = 5

2 + 5 = 7   9 - 5 = 4   +

-3   -3   3 + 2 = 5

=   =

2 + 4 = 6

How many of each shape are there?

◯ _6_   ▢ _5_   △ _5_

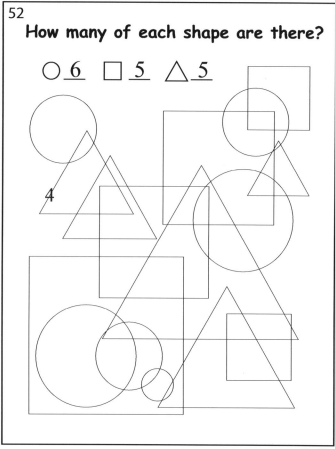

Solve this!

$$20 + 20 + 20 = 60$$

$$20 + 5 + 5 = 30$$

$$5 - 2 = 3$$

$$1 + 20 + 5 = \boxed{26}$$
?

CONGRATULATIONS!

You're truly amazing! I am sure there are some obstacles along the way; it was great you persisted through and finished the job!

If you want to continue with more activities, just send me an email to support@kidsactivitybooks.org and I will send you some for free.

My name is Jennifer Trace and I hope you found this workbook helpful and fun. If you have any suggestions about how to improve this book, changes to make or how to make it more useful, please let me know.

If you like this book, would you be so kind and leave me a review on Amazon.

Thank you very much!
Jennifer Trace

Congratulations
Logic Puzzle Star:

THE BEST!

_____

Date:_____   **Signed:**_____

Printed in Great Britain
by Amazon